I0471924

This report is part of a series of reports on technical rescue incidents across the United States. Technical rescue has become increasingly recognized as an important element in integrated emergency response. Technical rescue generally includes the following rescue disciplines: confined space rescue, rope rescue, Trench/colapse rescue, ice/water rescue, and agricultural arid industrial rescue. The intent of these reports is to share information about recent technical rescue incidents with rescuers across the country. The investigation reports, such its this one, provide detailed information about the magnitude and nature of the incidents themselves; how the response to the incidents was carried out and managed; the impact of these incidents on emergency rcsponders and the emergency response systems in the community; and the lessons learned. The U.S. Fire Administration greatly appreciates the cooperation and information it has received from the fire service, county and state officials, and other emergency responders while preparing these reports.

This report was produced under contract EMW-94-C-4436. Any opinions, findings, conclusions, or recommendations expressed in this publication do not necessarily reflect the views of the U.S. Fire Administration or the Federal Emergency Management Agency.

Additional copies of this report can be ordered from the U.S. Fire Administration, 16825 South Seton Avenue, Emmitsburg, MD 21727.

Search and Rescue Operations
Following the Northridge Earthquake
Los Angeles, California
January 17, 1994

Local Contacts:

Wayne Ibers
Los Angeles County Fire Department
1320 North Eastern Avenue
Los Angeles, CA 90063-3294

Keno Devarney
Los Angeles County'Fire Department
1320 North Eastern Avenue
Los Angeles, CA 90063-3294

Gary Siedel
Los Angeles City Fire Department
227 North Lake Street
Los Angeles, CA 90026

Larry Collins
Los Angeles County Fire Department
1320 North Eastern Avenue
Los Angeles, CA 90063-3294

OVERVIEW

At 0431 hours on January 17, 1994, a 6.7 magnitude earthquake struck the Los Angeles area. The epicenter was located within the Northridge area of the San Fernando Valley, approximately 20 miles east of downtown Los Angeles (see Appendix A). Many experts believe that because the earthquake occurred on a holiday morning, casualties were significantly lower than they would have been if the quake had happened at any other time. Over the following days, several aftershocks occurred that continued to damage structures as well as jeopardize the safety of rescue personnel.

For firefighters and other emergency responders, the Northridge Earthquake was another in a long line of disasters that have challenged their resolve in recent years. Southern California has been walloped by major wildland fires, floods, mud slides, earthquakes, and riots. This quake marked the third time in three months and the fourth time in two years that thousands of firefighters from California and other states responded to provide mutual aid for a major disaster in Los Angeles County. [1]

Sadly, 57 people lost their lives as a result of the quake, thousands were injured, and tens of thousands were left homeless. More than 100 major alarm tires occurred. As bad as it was, this was not the fabled "Big One" for Southern California. Nor is it the last time a large earthquake will hit densely populated areas of L.A. County. For fire and rescue services everywhere, the Northridge Earthquake should serve as a wake-up call to prepare in earnest for major disasters.[2]

In addition to managing its own impacted areas (more than 2,500 square miles and 50 cities), the L.A. County Fire Department deployed its 56-person FEMA USAR Task Force and other resources to assist the L.A. City Fire Department and helped manage mutual aid resources from across the nation.[3] The following report is an account of the initial response by rescue personnel and the subsequent hazard mitigation by agencies involved.

Technical Rescue Incident Investigation Project Explained

The Technical Rescue Incident Investigation Project is an effort of the US. Fire Administration to document case studies of certain technical rescue incidents. The project seeks to produce documentation in a "lessons learned" format in order to provide local emergency responders, trainers, federal and state agencies, and other interested groups enhanced knowledge about technical rescue response and safety.

History of Earthquake Preparedness

Because of the history of earthquakes and related seismic activity in Southern California, emergency service personnel as well as citizens are acutely aware of the potential for the "Big One." Over the past several years, public safety agencies have taken a lead role in organizing the community to prepare for a major disaster. The California Office of Emergency Services, City and

---

[1] Excerpt from *Rescue Magazine,* May/June 1994.
[2] Ibid.
[3] Ibid.

County Fire Departments, City Police, County Sheriffs Office, public works departments, utility companies, the Red Cross, and area hospitals have all played a role in community disaster preparedness. Each of these entities is assigned a role in a disaster response. The local fire departments have hundreds of thousands of hours training first responders for their role during a disaster such as the Northridge Earthquake. The Los Angeles City and County Fire Departments spent hundreds of thousands of dollars on the necessary equipment for handling structural collapse incidents. Although the equipment purchases and training of fire department personnel are not complete as of this time, both agencies have achieved a large portion of their goals in the overall preparation for a large regional disaster.

As part of the preparation, agencies with L.A. County developed a plan for immediate response after a disaster occurs. The EEOP, or Earthquake Emergency Operational Plan, is intended to provide a framework for command, control, and management of an earthquake disaster. As part of the EEOP, immediately after a disaster strikes, departments activate Earthquake Emergency Mode (EEM). This mode of operation provides specific plans under which individual companies are to operate. Part of this plan calls for the immediate recall of off-duty personnel, communications checks, and a drive-through assessment of damages and hazards.

All of the preplanning and preparation allowed for a timely response of emergency services personnel to the disaster sites. This preparation accounted for many lives being saved which would have otherwise perished in the disaster. The community response preplanning also saved millions of dollars in damage to structures as well as the infra-structure in the Los Angeles area. There is no doubt in anyone's mind that the investments in preplanning and disaster preparedness yielded exponential returns in terms of lives and property saved.

Acknowledgments

The USFA would like to acknowledge the cooperation of the Los Angeles County Fire Department and the City of Los Angeles Fire Department and their representatives in providing details of the initial response and subsequent rescue operation. Their detailed analyses were crucial in providing the specific "lessons learned" to be shared with all first responders. We would also like to thank *Fire Engineering Magazine* (8/94) for its detailed reporting of events after the fact.

I.   Initial Response

Because the incident occurred at 0431 hours, most on-duty firefighters were in quarters and in bed. There were many reports of firefighters waking up after being thrown out of bed by the earthquake. Just about all personnel quickly realized that an earthquake had occurred and proceeded to their designated assignments. As part of the EEM, bay doors were raised and all station apparatus was driven out onto the aprons. Because of damage to some stations, this proved to be very difficult. In many cases, apparatus doors were jammed shut because of structural damage and required forcible exit.[4] As part of the plan, station officers first assured the safety of their personnel and then surveyed the damage to their physical resources (station, apparatus, equipment, etc.). A radio check provided Battalion Commanders with a quick assessment of the conditions of human and physical resources.

As determined by the pre-plan, companies performed a drive-through assessment of their first due area to determine the extent of the damage and the resources needed. The drive-through allowed area commanders to identify where the worst damages occurred and allocate resources accordingly.

As fire companies began driving through their districts, it was clear there was heavy damage and potential for trapped victims in the San Fernando and Santa Clarita valleys and parts of the L.A. basin. Apartment buildings, freeway overpasses, and large commercial buildings had collapsed throughout these areas.[5]

As can be expected, fire companies had to stop during their drive-throughs for fires and the immediate threat to lives. In some cases, after removing all the occupants, fire companies had to let structures burn due to the lack of water and the disruption of water flow in hydrants. Although this tactic proved beneficial to the people whose lives were saved, it slowed the damage assessment and resource allocation process. Company commanders were forced into a basic triage decision of what to save and what to let burn. Understandably, these decisions were very difficult for those who made them. Staying to the predetermined "game plan" was a critical component for the overall success of the operation. With early initial recon reports, the Command Staff was able to allocate

---

[4] Ibid.

[5] Ibid.

necessary resources to the areas with the most need. Although there were literally hundreds of isolated fires caused by the earthquake, search and rescue operations were confined to a relatively small area.

At the same time fire companies were performing drive&roughs, the air operations units were busy implementing their own specific disasters plans. Helicopters, staffed with on-duty personnel, quickly became airborne. Off-duty aviators were immediately recalled so that the maximum number of aircraft could be launched in an attempt to identify the scope of the problem. With pre-designated plans, aerial reconnaissance of the area began immediately. The scene from the air above the San Fernando and Santa Clarita valleys was eerie. Dozens of major fires were burning, with more smoke rising in the distance. Entire blocks of homes were on fire in several locations. At sunrise, a pall of smoke gathered in heavy layers above the San Fernando Valley. With water mains severed and gas mains shooting flames from the streets in many regions, firefighters on the ground were having a tough time holding their own. There was a constant stream of radio traffic as engines and strike teams called for more water and attempted to set up water relay operations from the few working hydrants to be found.[6]

The gravest concern, however, was the condition of the dams holding back the water reservoirs above Los Angeles. It has been estimated that one dam failure alone could kill 20,000 people. With over ten of these dams in the L.A. area, they were first to be assessed for structural damage, followed closely by hospitals, industrial facilities, fuel storage facilities, and freeways.

While the Command Staff was making preparations for the control and command of a large campaign operation, fire companies and Battalion Commanders organized task level operations. Many lives were saved by the quick thinking and heroic actions of firefighters. The ingenuity of firefighters allowed for the effective use of damaged resources during the incident. In close communication with DWP (Department of Water & Power), fire company personnel were able to identify which hydrants and mains were useable and which were not. Around the clock relay pumping operations were placed in service to provide a continuous flow of water for firefighting operations.

---

[6] Ibid.

5

With most recon reports in, emergency resources and attention were directed to the Northridge area for search and rescue operations. The quick, aggressive, and well-coordinated response in the initial phase of this disaster saved many lives, saved millions of dollars, and allowed Commanders to get a handle on the problem very quickly.

II.   Search and Rescue Operations

Shortly after the initial insult, emergency services personnel realized that search and rescue operations were going to be concentrated in a very small area. The immediate area surrounding California State Northridge was the hardest hit. Rescuers encountered three major rescues and several potentially hazardous buildings. Each of the three major problem structures will be examined individually for a close look at lessons learned.

Science Lab at Cal-State Northridge

As a result of earthquake damage, fires broke out in three science buildings at Cal-State Northridge. It is still undetermined as to the exact cause and origin of these fires. Because of the large volume and quantities of chemicals used in higher education laboratories, some theorize that the knocking over and subsequent mixing of chemicals caused the spontaneous combustion of materials. Regardless of the cause, the three buildings with a haz-mat placard (labeled 4x4x4) had well-advanced fires within. The three-story masonry structures sustained light structural damage. Due to the unique nature of higher education research labs, there is always a possibility of an instructor or student conducting research after normal working hours, therefore the possibility of life hazards existed.

While conducting search and rescue operations in the multi-story structures, fire companies realized the potential for secondary structural collapses due to aftershocks and fire involvement in the structures. Companies were committed early to the interior of the structures before the haz-mat involvement potential was fully realized. After the 4x4x4 was taken into consideration, companies were pulled out for a thorough analysis of the rescue operation's risks and benefits. By then, students, teachers, and maintenance personnel were on scene for further consultation of the haz-mat component. Attempts were made to gather MSDSs and identify specific products in exact locations while companies worked to contain the fire from a defensive position.

After consulting with the Cal-State Northridge staff and completing a haz-mat size-up, command developed a plan that would send crews back to the interior for search and rescue and firefighting operations. A hot zone was established based on atmosphere readings and wind conditions, and crews resumed operations. With specifically directed positive pressure ventilation, interior crews made a well-coordinated aggressive interior attack on the fire. The attack stopped the forward progress of the fire into areas containing dangerous hazardous materials. This tactic proved to be crucial in minimizing the exposure potential of personnel operating on the scene.

Lessons Learned

Many individuals gained valuable experience in dealing with the Cal-State Northridge Labs. There are many individual lessons learned because of this experience. A few in particular stand out as worthy of consideration if you are faced with a similar situation.

- Initial crews initiated an aggressive interior attack on the Building 2 fire at the Northridge Labs. Crews engaged in structural firefighting on a partially collapsed multi-story structure, with a low probability of trapped victims. These interior operations were conducted while aftershocks continued to occur. It was not until the explosions began that firefighters realized the haz-mat potential in the building and retreated to an exterior defensive position. The move was essential since it allowed time for a thorough analysis of the risks and benefits of interior firefighting operations. After consulting with the Cal-State Northridge staff and technicians as to quantity and location of hazardous materials, a plan to halt the forward progress of the fire was formulated and implemented. The benefits of a well-coordinated interior attack were apparent to all firefighters on the scene.

- Because of several broken underground water mains in the area, crews did not have an adequate water supply for firefighting operations, and were forced to improvise. Firefighters located water sources (swimming pools) in the immediate vicinity and began drafting operations. Because of the distance involved, a tandem operation was employed that provided just enough water to carry out firefighting operations. With the judicious flow of water through master streams and hand lines, crews were able to make do with the situation and bring the fire under control.

- On scene Commanders consulted with faculty staff to establish the hazard potential and an action plan. Based on information provided by the staff, a very specific and coordinated plan was developed. The mutual agreement about the best approach to hazard mitigation proved invaluable to firefighting forces.

- Research of the involved products and subsequent atmospheric monitoring established exposure potential. Personnel who were directly involved with interior operations were screened and their personal protective equipment was surveyed for contamination levels. Documentation was begun to establish line of duty injury or death if it ever becomes necessary. In the heat of a huge regional disaster, the on-scene commander was able to make decisions that provided for the continued health and safety of his personnel.

The Northridge Mall Parking Garage

Approximately one and a half hours after the initial tremors were felt, a security guard at the Northridge Fashion Mall called in a report of a person trapped in the parking garage. The first responding crew sized up the situation and called for additional resources. A lone individual was operating a street sweeper on the first floor of a three-story parking garage. The structure, a reinforced concrete parking garage, had collapsed, crushing Salvadore Pena in his street sweeper. For the next several hours, rescuers worked hard while consciously risking their lives in trying to free Mr. Pena.

Because the vehicle he was in was crushed, the fuel tank ruptured spewing gasoline onto the first floor. An ignition source was all that was needed to finish the victim. After locating the victim, the Captain of the first arriving unit called for the application of 60 percent foam to suppress the flammable vapors and reduce the hazards of the upcoming rescue attempt. As additional units responded to the scene, a discussion ensued about what resources were necessary to conduct the extended operation. An L.A. City Fire Department USAR unit was dispatched to the scene. On arrival, the Captain made a triage decision that more lives could be saved with his personnel and equipment if they were used at the Meadows Apartments operation. That USAR crew left the parking garage and responded to the Meadows Apartments, after calling for mutual aid from L.A. County Fire Department.

8

L.A. County Fire Department's USAR-1 responded to the scene. Captain Wayne Ibers was put in charge of the rescue operation. After a thorough size-up of the situation, USAR-1's crew developed a rescue plan for Salvadore Pena. The three-phase plan was initiated after a briefing on rescuer safety. The safety briefing discussed potential hazards, emergency escape plans, lifting and cribbing operations, and communication procedures. All involved members agreed on a rescue strategy, and crews began Phase 1 of the operation.

The strategy of Phase 1 was to remove as much weight as possible from the second and third floors off the victim. This was done by jackhammering and removing as much concrete as necessary to access the victim on the first floor. The opening from above the victim had to be large enough to allow for the movement of personnel and equipment down to the first floor. From the onset, it was decided that at lease one paramedic would stay with the victim to continually monitor his condition, and to assure him that all that could be done was being done. Many times throughout Phase 1, the operation was stopped because of aftershocks. Each aftershock brought reminders to all personnel of the dangers of working inside a collapsed structure.

Throughout Phase 1 members were rotated to provide rehab services. Periodically, Captain Ibers and others reviewed the merits of the rescue plan and adjusted them accordingly. Finally, after a few hours of intense work, rescuers had relatively good access to the victim. A break was taken and command held a briefing on the specifics of Phase 2. The plan was to lift the heavy reinforced concrete beams off the street sweeper so that extrication of the victim could be completed.

As Phase 2 was implemented, crews provided continued vapor suppression with 6-percent foam and positive pressure ventilation to the immediate rescue area. A timber (cribbing) cutting station was set up a short distance away from the rescue site. This was done to minimize the vibration and the noise. As crews worked on the first floor, they would call out dimensions of lumber to be cut by the cutting team. The rescuer's approach was to alternately lift the heavy beams with high pressure air bags and crib as they went along. Each lift was only a few inches at a time, while cribbing and wedges prevented the beam from falling back down. Although this was a very difficult and delicate task, crews did an excellent job of lifting and cribbing in the tight space.

During the lifting of the beams, paramedics monitored the patient's condition and watched for signs of the "crush syndrome." The beams were lifted approximately twelve inches off the sweeper. This allowed Phase 3 of the operation to begin, which would involve a well orchestrated attempt to extricate the victim from the twisted wreckage of the sweeper. Because the collapse crushed the sweeper around the victim, it was very difficult to cut away parts of the vehicle to access the victim. Once the extrication was complete, the victim was properly packaged for removal from a collapsed structure. Within a few minutes, Mr. Pena was loaded into the helicopter and transported to the hospital. After nine hours of being buried under a rubble pile, the victim was finally free.

Captain Ibers sent all personnel to rehab after the victim had been transported to the hospital. After rescuers were sufficiently recuperated, Captain Ibers reviewed the operational plan for removing the air bag from below. Rescuers were reminded of the importance of personal safety and to not let their guard down during the equipment removal phase of a rescue. Because other rescues in the area were still being conducted, there was no time on the scene to do a critique of the operation. Members of all the companies involved gathered their equipment and returned to service. An outstanding job was done by all members involved in the rescue of Salvadore Pena.

Lessons Learned

- The standardized USAR training given to members enabled three different agencies to work together for a smooth operation.

- A rescue plan was discussed and agreed upon by participating rescuers. Each member clearly understood the plan and his/her individual role. The plan was continually updated and refined, based on existing conditions.

- Rescuers understood that an operation in a confined space had the potential for an explosion. Vapor suppression was maintained, as well as positive pressure ventilation.

- Pre-established safe zones and evacuation routes were identified in case of secondary collapse. They were necessary on several occasions.

Bulky, structural firefighting helmets are improper for rescue operations. Construction-style helmets are preferable.

Lifting and cribbing operations were executed methodically, a little bit at a time, with time allowed between lifts to see how the structure was affected.

Command as well as company officers continued to evaluate the need for rehab of personnel,

Proper personal protective equipment was used for personnel operating in the hole. Jumpsuits worked much better than turnouts. Knee pads and eye protection were necessary for the comfort and safety of rescue personnel.

Color-coordinated hose lines were used for simultaneous lift bag operations to ensure proper coordination.

Clean up edges of tunneling operation in reinforced concrete structures. Rescuers cut rebar and wire as close to concrete as possible to eliminate hazards.

Experienced rescuers operating on the scene could feel the difference between small magnitude aftershocks and big ones. When the big ones were felt, they evacuated the hole; when the little ones hit, they kept on working.

The Northridge   Meadows Apartments

The Northridge Apartments was a three-story wood frame stucco exterior construction apartment building. The initial earthquake caused the building to collapse. Because of the first floor "soft story" construction, the second and third floors "pancaked" down onto the first story apartments. On the initial drive-through by LAFD units, it was not even noticed that there had been a collapse of the building. The complete collapse of the first floor made the structure look like a two-story building. When first responders finally did realize there had been a collapse and many victims were involved, they requested additional resources and began rescue operations.

Initial efforts by first responding crews were aimed at assisting victims (walking wounded) out of the building. Crew members used the hailing method to notify and attempt to locate the victims. Crews were faced with the difficulty of dealing with many untrained, well-intentioned survivors trying to reach loved ones and assisting where they could. A quick search and subsequent rescue saved many lives. This was accomplished without the use of specialized USAR equipment and personnel. Shortly thereafter, Chief Robert DeFeo took command of the incident.

Because of the time of day and the type of occupancy, Chief DeFeo realized that there was a possibility that many victims were still trapped in the collapse. Immediately, an Incident Command System (ICS) was put in place that would properly support the campaign operation. The complex was divided into three divisions for a manageable span of control. Division Commanders were briefed on "the plan" and began a more thorough search of the collapsed structure. Crews were still operating without the use of specialized search and rescue equipment. After the second and third stories were cleared, crews worked to gain access to the first floor.

Accessing the first floor was difficult. The deck between the first and second floors consisted of 2x12 floor joists, one inch plywood decking, and two inches of lightweight concrete. Initially, crews attempted to cut through the concrete with rescue saws. This created the problems of dust, noise, and vibration. Rescuers soon abandoned those efforts and used picks and sledge hammers to break up the concrete. After enough concrete was removed, the plywood decking and floor joists were cut with chain saws. In many apartments, rescuers realized that the second and third floors had shifted during the initial collapse by as much as ten feet. It soon became apparent that apartments were not lined up vertically. Some attempts to cut through the second story bedroom floor did not yield entry into the first story bedroom as they initially thought it would. As rescuers found dead victims, they were instructed to cover the bodies and leave them in place as per agreements with the L.A. County Coroner's Office.

During this phase of the operation, first responders rescued several trapped victims. By this time, Command had established a medical group to treat and transport victims. Command had also requested a PIO (public information officer) to deal with the media. LAPD had established scene security at the request of the Incident Commander. A primary search of the apartments had found fourteen Level 4 victims. LAPD USAR-1 was on the scene by now, assisting crews in rescuing the remaining live victims. Specialized equipment, such as high pressure air bags, were used to lift parts of the collapsed structure off some of the victims. LAPD K-9 units were able to assist in searching for victims. The K-9 search was only marginally successful.

By late afternoon, mutual aid from L.A. County (California Task Force 2) arrived on the scene. CATF-2, with a full complement of equipment and personnel, was dressed and ready to begin operations. At this time, crews from LAFD had been working continuously for twelve hours. Chief DeFeo started decommitting LAFD crews and briefed CATF-2 Leader, Chief Keno Devamey, on what had been done and what was left to do. Chief DeFeo also provided a detailed map of the entire apartment complex. Chief Devarney assembled all of his team managers and formulated a specific plan for a secondary search of the complex. The plan was similar to what LAFD had done, in that they were to access the first floor vertically from the bottom of the second floor apartments.

Search and rescue teams from CATF-2 worked closely to open up potential void spaces. One of the benefits of the USAR cache of equipment was the availability of technical search equipment, such as a fiber optic SearchCam. This allowed team members to look into rooms without having to make an opening large enough to get a rescuer through. At one point during the secondary search of the building the power was restored to the area. Because the disconnect to the building had not been "pulled," power in the interior of some apartments came on. This caused several working fires in some apartments and LAFD units had to be called to the scene for firefighting operations. After the fires were extinguished and the power to the building had been cut, CATF-2 went back to work to get a secondary "All Clear."

Because of the size of the complex and the time it would take one task force to get an "All Clear" in the entire complex, an additional FEMA Task Force was requested by Command. CATF-6 from Riverside was on the scene at approximately 2230 hours. The plan at that time was to split the complex in two, while CATF-2 took one end of the complex, and CATF-6 took the other. Crews continued the arduous search of the complex and by 0330 hours on the 18th of January they had an "All Clear" of the entire structure. Almost 24 hours after the earthquake hit, the entire 163 unit complex was searched and cleared of all victims. The final toll was sixteen dead, with numerous injured. The preparation and inter-agency coordination paid off in saving many lives.

Overhead View of the Northridge Meadows Apartments

N

Key Rescues ✓

104 Robert Dorsey	106 Steve Langdon
106 Jerry Prezioso	110 Alan Hemsath

RESEDA BLVD.

ALLEY

COURTYARD

ALLEY

PARKING STRUCTURE

103† 203 303 | 101† 201 301 | 205 305 | 104★ 207 307 | 106✓ 209 309

204 304 | 102† 202 302 | 206 306 | 105† 208 308 | 210 310

266 366 | 138 265 365 | 110✓ 218 318 | 108† 216 316 | 212 312

264 364 | 137 263 363 | 111† 219 319 | 109 217 317 | 214 314

262 362 | 136 261 361 | | 107 215 315

260 360 | 135 259 359 | 221 321 | 112 220 320

258 358 | 134 257 357 | 223 323 | 114 222 322

256 356 | 132 255 355 | 225 325 | 115 224 324

254 354 | 131 253 353 | 227 327 | 116 226 326

252 352 | 130 251 351 | | 117 228 328

250 350 | 129 249 349 | 121 232 332 | 119 230 330 | 118 229 329

248 348 | 128 247 347 | 122 233 333 | 122 233 333 | 123† 234 334

246 346 | 243 343 | 127† 241 341 | 126 239 339 | 124† 235 335

245 345 | 244 344 | 242 342 | 240 340 | 238 338 | 237 337 | 125 236 336

Reproduced with permission from Rescue Magazine

Lessons Learned

- Early implementation of an ICS that would support a large multi-agency operation allowed for the organization of efforts and safety of personnel operating on the scene.

- Cooperation with the building manager helped rescue personnel obtain a specific layout of the apartment complex and individual units so that the rescuers could search the high probability areas (i.e., bedrooms) first.

- Utilities must be secured as early as possible to ensure the safety of rescuers.

- Rescue operations are more efficient when command manages untrained civilians and residents to account for trapped victims and reduces secondary injuries caused by working in the collapsed areas.

- Establishing an EMS triage and staging area improved the efficiency of caring for the sick and injured.

- After initial rescue efforts by first responders, a specific plan for search and rescue was implemented, which reduced freelancing and improved firefighter safety.

- In wood frame construction, basic hand tools proved more valuable than specialized equipment.

- Basic truck company operations in opening up the collapsed apartments proved very effective and saved several lives.

- Inter-agency communication and cooperation working under one IC was very effective.

- Standardized ICS and Structural Collapse Training (RS I and RS II) made for very safe and efficient operations.

It Could Have Been Worse

The Northridge Earthquake might have been far more destructive under the following conditions:

1. If the quake had occurred at a different time on a different day:
If the Northridge Quake had occurred at 4:00 in the afternoon of a normal school/workday, the damage that occurred would have been devastating.

- Collapsed freeway overpasses would have crushed many more automobiles. These rescues are particularly dangerous and time-consuming because they require breaching and lifting tons of unstable concrete slabs. After removing such debris, rescuers still must extricate victims from wrecked autos.

- Vehicle accidents on open roads and freeways would have been a major source of casualties.

- Structures such as the Kaiser Permanente building, Northridge Fashion Center, and various other commercial and office buildings that suffered major collapses would have been full of people during normal business hours, resulting in hundreds of fatalities. Rescue operations to remove live victims might have taken more than a week.

- Parking structures would have been full of automobiles and people. These incidents might have required days to complete rescue operations.

2. If the quake had been stronger:
The Northridge Quake pushed many buildings to the limits of their integrity. The vertical motion caused by this thrust fault was surprisingly strong, but stronger shaking would almost certainly have brought more buildings down.

3. If the quake had lasted longer:
The Northridge quake only lasted 15 seconds. Even without an increase in magnitude, many buildings would have collapsed if the shaking had lasted much longer. Some buildings were on the verge of collapse when the shaking stopped.

4. **If the quake had occurred in a different location:**

There are worse places to have a 6.8 earthquake. Some areas closer to the coastline have higher degrees of liquefaction danger. Many of these areas are densely populated and have thousands of buildings that would not perform well in a major earthquake.

5. **If the quake had occurred during major fires, floods, or other emergency conditions:**

Los Angeles area firefighters and other emergency responders have faced several major disasters in recent years. These include the floods of 1992 and 1993; the 1992 Los Angeles riots; and the fire storms in the fall of 1993. L.A. County required the aid of thousands of firefighters from other cities, counties, and states during these incidents. More than 900 fire engines from across the state responded to the Malibu fires alone. If the Northridge Quake had occurred at the height of any of these events, California's fire and rescue resources would have been severely taxed.[7]

---

[7] "It Could Have Been Worse," *Rescue Magazine* May/June 1994.

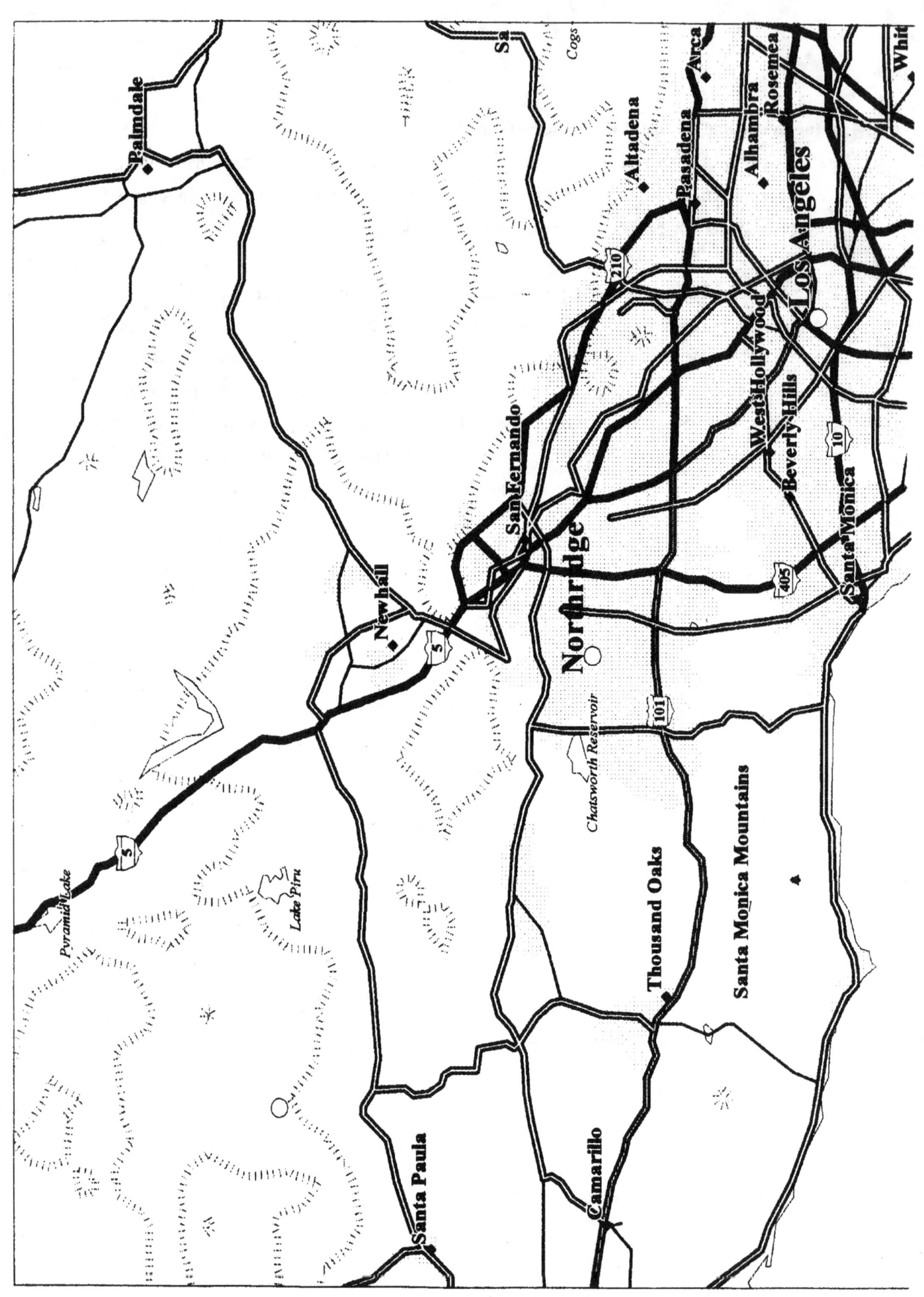

Appendix B

Photographs of Earthquake Damage

Initially firefighters thought that the Northridge Meadows Apartment building was a two-story structure. The building on the right is the original height of all the apartments.

Example of shoring techniques used by the LAFD and the LACFD to prevent a secondary collapse of the high-rise hotel parking garage.

Collapsed portion of Northridge Meadows Apartments. Due to the soft story construction, the structure not only collapsed, but shifted 10 feet.

What remained of cars located in the parking garage below the Northridge Meadows Apartments.

*U S Government   Printing   Office:1996-719-595/82752

www.ingramcontent.com/pod-product-compliance
Lightning Source LLC
Chambersburg PA
CBHW081422170526
45166CB00010B/3437